E
G Goor, Ron
    Signs

DATE DUE

MAY 21 '85	OCT 26 '85	APR 14 '86	DEC 01 1986
MAY 30 '85	NOV 02 '85	MAY 08 '86	DEC 08 1986
JUN 24 '85	NOV 19 '85	MAY 27 '86	JAN 24 1987
JUL 01 '85	DEC 30 '85	JUN 14 '86	FEB 07 1987
JUL 17 '85	JAN 17 '86	JUN 30 '86	FEB 18 1987
JUL 31 '85	FEB 06 '86	JUL 02 '86	MAR 23 1987
AUG 10 '85	MAR 01 '86		JUN 17 1992
AUG 25 '85	MAR		1992
SEP 07 '85	MAR		10 1992
SEP 24 '85			
OCT 09 '85	MAR 26		

PRESS CARD

  FB

HARRIS COUNTY PUBLIC LIBRARY

HOUSTON, TEXAS

SIGNS

Ron and Nancy Goor

Thomas Y. Crowell / New York

Signs
Copyright ©1983 by Ron and Nancy Goor
All rights reserved.
Printed in the United States of America.

Library of Congress Cataloging in Publication Data
Goor, Ron.
  Signs.

  Summary: Text and photographs present
familiar signs met in daily life.
  1. Traffic signs and signals—Juvenile
literature. 2. Signs and signboards—Juvenile
literature. [1. Signs and signboards.
2. Traffic signs and signals] I. Goor, Nancy.
II. Title.
TE228.G66 1983   625.7′94   83-47649
ISBN 0-690-04354-6
ISBN 0-690-04355-4 (lib. bdg.)

Designed by Al Cetta
1 2 3 4 5 6 7 8 9 10
First Edition

For our son Danny,
who gave us the idea
for this book

Signs...

signs…

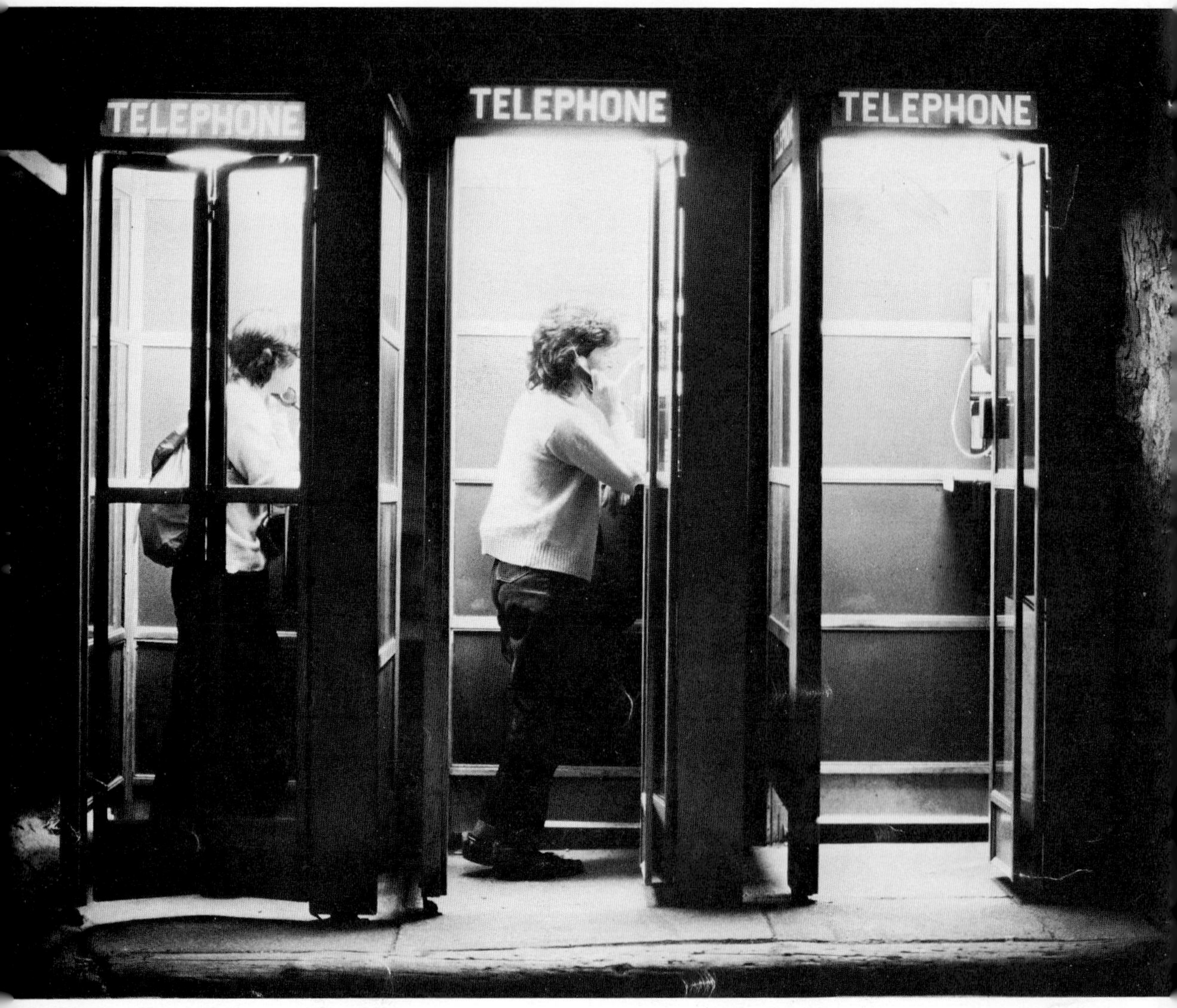
TELEPHONE
TELEPHONE
TELEPHONE

signs are everywhere.

You read signs

when you go to school,

when you cross the street,

when you ride your bike,

and when you take a bus.

Signs are useful.

They tell you where to go.

Signs tell you where *not* to go.

Signs tell you what to do.

Signs warn you to be careful.

Signs tell you where to find help.

What do these signs tell you?

Match the picture with the sign.

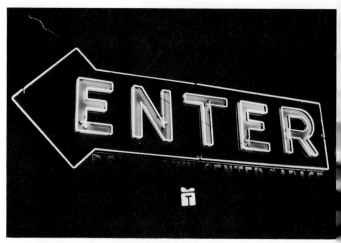

Match the signs that mean the opposite.

Everywhere you go,

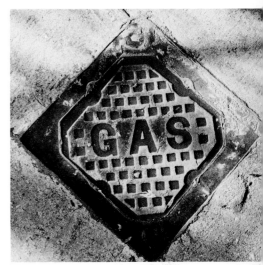

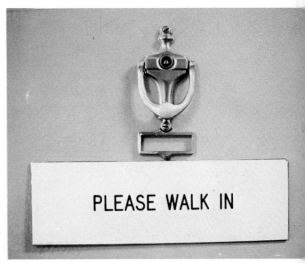

you can read signs.

Harris County Public Library
Houston, Texas